THE INTERNATIONAL SPACE STATION

Written by
Clive Gifford

Illustrated by
Dan Schlitzkus

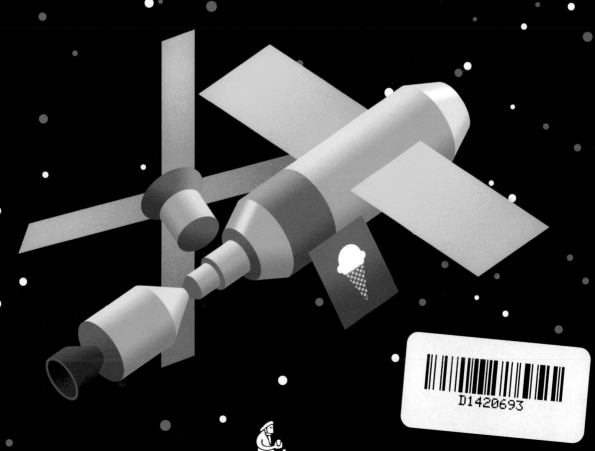

WAYLAND

First published in 2017 by Wayland
Copyright © Hodder and Stoughton 2017

Wayland
Carmelite House
50 Victoria Embankment
London
EC4Y 0DZ

Editor: Victoria Brooker
Designer: Krina Patel

A cataloguing record for this title is available at the British Library.

ISBN: 978 1 5263 0216 8

Printed in China

Wayland, part of Hachette Children's Group
and published by Hodder and Stoughton Limited
www.hachette.co.uk

Contents

What is a Space Station?

Space stations are craft which stay in space for long periods of time. They travel on a path around Earth called an orbit. In 1971, the Soviet Union launched the world's first space station, Salyut 1. It spent 175 days in space, orbiting Earth over 3,000 times.

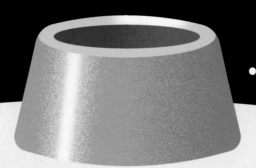

Salyut 1

Home In Space

Space stations provide astronauts with places to live and work, sometimes for missions lasting months at a time.

On board, the crew work hard, carrying out lots of experiments to discover how space affects materials, humans and other living things.

Skylab

America's first space station, Skylab, was launched in 1973 by NASA. It was the first to contain a shower, and a freezer to offer the first ice cream in space! Skylab was abandoned in 1974 but stayed in space until 1979 when it fell back to Earth.

The craft was destroyed when it re-entered the Earth's atmosphere and small pieces of it fell into the Pacific Ocean and across parts of Western Australia. Officials in the small Australian town of Esperance sent NASA a US$400 fine.

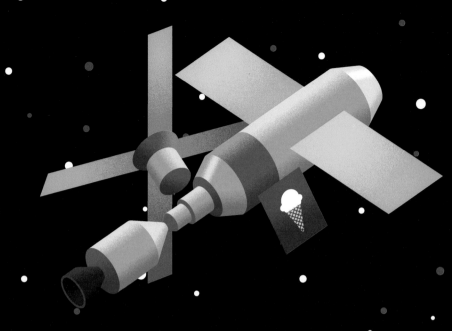

Mir

Mir was the first space station to be assembled from different sections, called modules, in space rather than launched from Earth as one complete craft.

It was designed to run for five years but in the end lasted 15 years and 31 days (1986–2001). During this time, it was visited by 104 different astronauts from 12 countries.

The International Space Station (ISS)

The biggest space station, the ISS, took more than ten years to construct. It is four times larger than MIR and five times larger than Skylab, and provides as much living space inside as found in a six-bedroom house. It typically holds a crew of six astronauts.

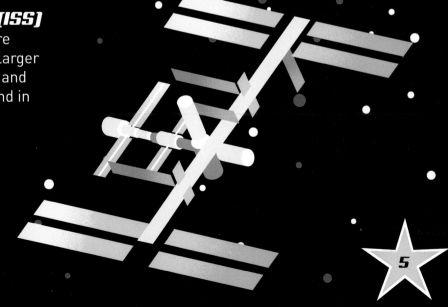

Space's Biggest Building Site

Building the ISS was a HUGE undertaking involving thousands of people. Five different space agencies were involved: NASA, Roscosmos (Russia), the Canadian Space Agency, JAXA (Japan) and the European Space Agency (ESA), which represents 22 different European countries.

Out In Space

More than 100 spaceflights were required to build the ISS. The first module was flown into space in 1998.

Astronauts have performed over 140 spacewalks (see pages 26–27) to help assemble the ISS. Before they travelled into space, astronauts practised every bit of the construction over and over again.

A NASA space shuttle docked with the early ISS when it was being constructed in space.

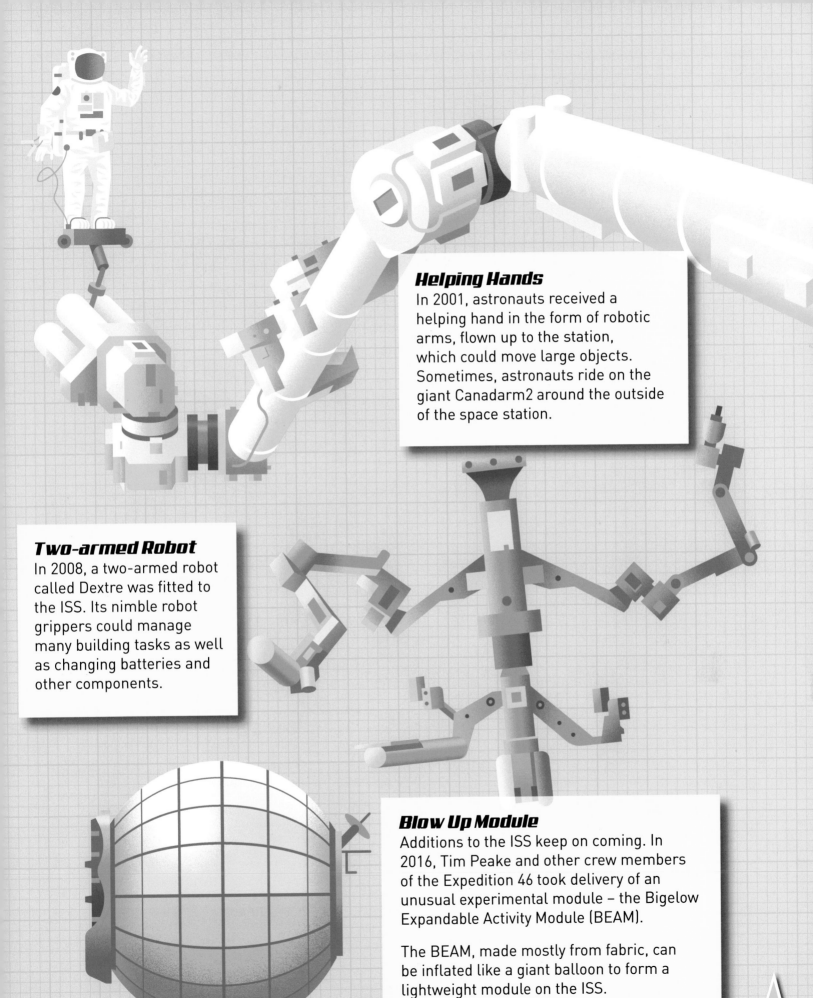

Helping Hands

In 2001, astronauts received a helping hand in the form of robotic arms, flown up to the station, which could move large objects. Sometimes, astronauts ride on the giant Canadarm2 around the outside of the space station.

Two-armed Robot

In 2008, a two-armed robot called Dextre was fitted to the ISS. Its nimble robot grippers could manage many building tasks as well as changing batteries and other components.

Blow Up Module

Additions to the ISS keep on coming. In 2016, Tim Peake and other crew members of the Expedition 46 took delivery of an unusual experimental module – the Bigelow Expandable Activity Module (BEAM).

The BEAM, made mostly from fabric, can be inflated like a giant balloon to form a lightweight module on the ISS.

Piece By Piece

This is it. This is how the International Space Station looks today. At almost 110 m long and 73 m wide, the space station on Earth would weigh 420,000 kg. There's no two ways around it, it's a whopper! From Earth, it is also the third brightest object in the sky, after the Sun and Moon.

Solar Array Wings (SAWs)
There are four pairs of these giant solar panels, measuring 73 m long. Each contains 32,800 solar cells which convert light from the Sun into electricity to provide the space station with power.

Main Truss
The main truss measures 109 m long and forms the station's backbone. Lots of her parts are connected to the truss.

Soyuz Spacecraft
One Soyuz spacecraft is always left docked to the ISS as a 'lifeboat' should there be an emergency and the crew need to leave suddenly.

In Orbit
The ISS orbits Earth at a distance of approximately 400 km above the planet's surface. In May 2016, the ISS celebrated performing its 100,000th complete orbit of Earth. That's a whopping 4,254,046,974 km, or almost 20 times the distance between Earth and Mars.

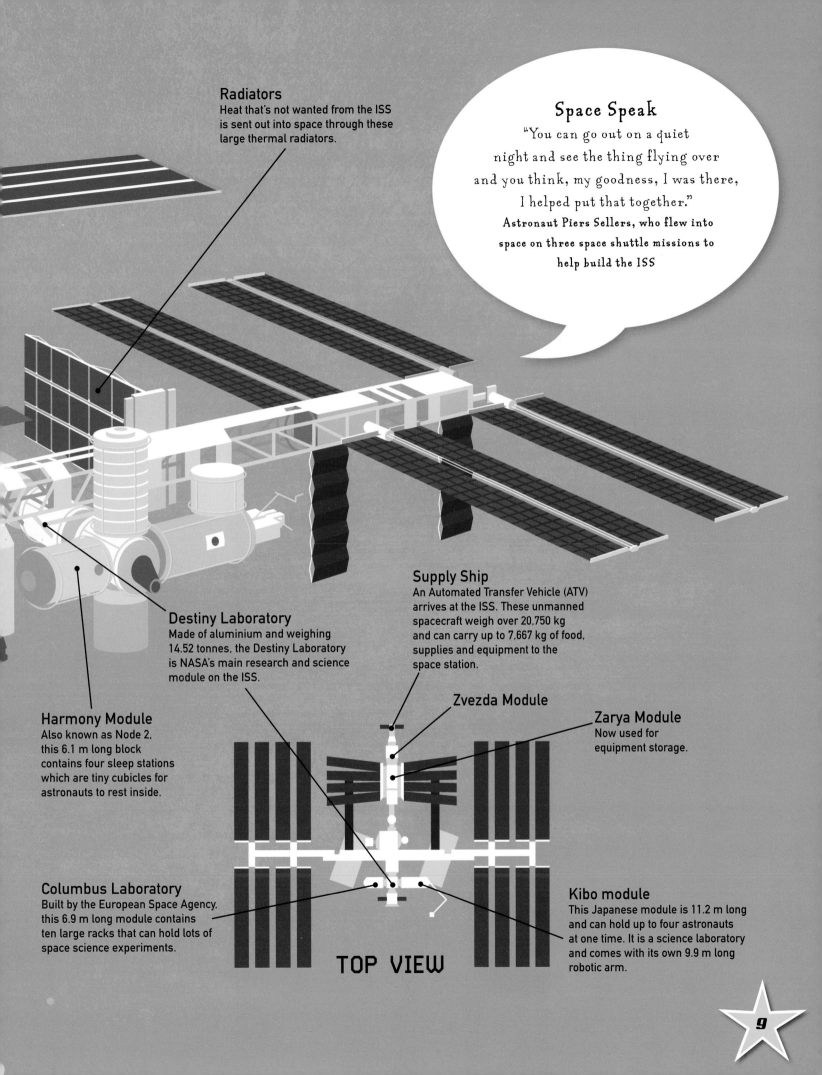

Radiators
Heat that's not wanted from the ISS is sent out into space through these large thermal radiators.

Space Speak
"You can go out on a quiet night and see the thing flying over and you think, my goodness, I was there, I helped put that together."
Astronaut Piers Sellers, who flew into space on three space shuttle missions to help build the ISS

Supply Ship
An Automated Transfer Vehicle (ATV) arrives at the ISS. These unmanned spacecraft weigh over 20,750 kg and can carry up to 7,667 kg of food, supplies and equipment to the space station.

Destiny Laboratory
Made of aluminium and weighing 14.52 tonnes, the Destiny Laboratory is NASA's main research and science module on the ISS.

Zvezda Module

Zarya Module
Now used for equipment storage.

Harmony Module
Also known as Node 2, this 6.1 m long block contains four sleep stations which are tiny cubicles for astronauts to rest inside.

Columbus Laboratory
Built by the European Space Agency, this 6.9 m long module contains ten large racks that can hold lots of space science experiments.

TOP VIEW

Kibo module
This Japanese module is 11.2 m long and can hold up to four astronauts at one time. It is a science laboratory and comes with its own 9.9 m long robotic arm.

So You Want To Be An Astronaut?

Over 220 men and women have visited the International Space Station. Some made flying visits as part of Space Shuttle or other crews. More than 100, though, have served long missions on the space station, typically lasting six months.

Big Demand

A lot of people want to head into space. When the European Space Agency announced they were looking for six astronauts in 2008, over 8,400 people applied.

Candidates with university degrees in science and engineering and experienced jet pilots with more than 1,000 hours flying time are the most likely applicants to succeed.

You also need to be the right age (usually between 27 and 46). No children have ever gone into space ... yet!

Train To Gain

Training is a full time job and can take three years or more. You have so much to learn from every aspect of how the ISS works to emergency first aid and CPR should a fellow astronaut become seriously ill.

In addition, trainee astronauts learn, practise and drill all the different skills required for life on the space station, from learning how to cook food to using the toilet in space.

Mind Your Language

Astronauts come from many different countries so learning how to speak not just English but the language of other crew members is vital.

With Russia providing many of the ISS modules, spacecraft and crew members, being able to speak and read Russian becomes a must!

сделать кнопку
не нажимать!

(Do not press button!)

держать дверь!

(Keep door shut!)

предупреждение!

(Warning!)

Vomit Comets

Astronauts get a taste of the weightlessness in space inside special planes, which fly sharply up and down lots of times.

The sudden shift in the plane's path creates short periods where astronauts float inside the aircraft. Some find it disorientating and feel sick, hence these planes' nickname: vomit comet.

Underwater Work

Another way of simulating the weightlessness of space is to place trainee astronauts underwater in giant tanks of water. Astronauts practise spacewalks and ISS repairs on a full-sized underwater replica of the space station.

Blast Off!

To get to the International Space Station is one wild ride. Astronauts sit inside a small spacecraft on top of a giant launch vehicle packed with fuel and many rocket engines. After a long countdown, the rocket engines fire and in less than nine minutes, the astronauts are in space!

Rocketing Away

The Soyuz FG is often used to send Soyuz spacecraft to the ISS. It stands 49.5 m tall and weighs 305,000 kg. In just 60 seconds, a Soyuz FG rocket can race from a standstill to 1,700 km/h. After 118 seconds, it can be travelling at over 5,400 km/h!

Stages To Space

Most launch vehicles are built with either two or three stages. Each stage has its own fuel, oxygen and rocket engines. When its fuel is used up, it separates from the rest of the launch vehicle and falls away. This reduces the rest of the vehicle's weight making it easier for the remaining rocket engines to propel the craft into space.

Feel The Force

Strapped into their seats, astronauts experience the launch first as a low rumble then the feeling of speeding up as the rocket blasts off skywards. The forces that occur during lift-off push the astronauts back into their seats with great force, around three times the gravity we experience on Earth.

Dock Stop

After all that high speed racing through space, the last part of the new crew's journey is conducted at a snail's pace. An automatic system called KURs uses computers and radar to gradually and gently ease the spacecraft into one of the ISS's docking ports which connect the two spacecraft.

In And Out

Once docked, there's several hours of safety checks before the astronauts leave their spacecraft and come on board the ISS. Astronauts enter and exit the space station through an airlock – a small compartment sealed by two doors. Only one door is open at any one time so that none of the air inside the ISS leaks out.

Life In Space

Space has major effects on the human body, especially if you spend a long time away from Earth. Most of these changes can be reversed in the first year that an astronaut spends back on Earth.

Which Way's Up?

An astronaut's sense of balance gets scrambled in space. Some astronauts feel sick and unsteady, especially in their first days on board the ISS.

Puffed Up

Astronauts' faces and necks go puffy from all the liquid in the body that is no longer forced downwards by gravity. This can lead to puffy faces and some problems with an astronaut's eyes. Those floating fluids can also stuff up your nose and impair your sense of smell ... but then, with no showers on board the ISS, that may be just as well!

Bones and Muscles

Bones lose a little of their strength and weight on long missions in space. They lose calcium, a substance that helps bone stay strong. On Earth, your muscles have to work hard to counteract gravity when you stand up, walk or lift and move objects. In space, the muscles have far less work to do and start to shrink and waste away. An astronaut could lose between a fifth and half of some muscles' strength if they do not exercise on a long mission.

Growth Spurt

Astronauts can grow up to 5–6 cm taller in space! On Earth, gravity presses down on the discs that separate the bones of the spine. On the ISS with its lack of gravity, the discs relax and expand, a bit like a spring that is no longer pressed down.

In The Gym

To combat the effects of microgravity on long space missions, astronauts hit the gym every day they're in space. They perform long workouts, averaging 120-150 minutes on special gym equipment developed to work well in space.

Strap In and Run

Astronauts go for a run on the ISS's T2 treadmill. They first have to strap themselves in using elastic straps like bungees fitted over their shoulders and round their waists. These pull them down onto the treadmill so that their feet strike the treadmill belt with a force similar to running on Earth.

A Day In The Life

No two days are precisely the same for the crew on the ISS as different experiments and work have to be performed and a variety of challenges faced. Astronauts may also be placed on slightly different timetables to each other. The timings are only approximate, but here are the details of a typical day on the ISS.

6.00: Wakey-Wakey!

Alarms and lights signal the start of the day. The space station runs on Greenwich Mean Time (GMT). This is the time zone used in the UK and has been chosen as it is roughly half way between the time zones of the two mission controls, one in Houston, USA and one in Moscow, Russia.

7.00-8.15: Breakfast Meeting

After a quick rub down with a soapy cloth, and sometimes medical checks and blood tests, the crew have breakfast. They connect with mission control back on Earth and go through their working day in detail so everyone knows precisely what they're doing that day.

8.15-12.30: Off To Work

The astronauts head off to work in the different laboratories and workspaces throughout the space station. Some crew members may be tasked with moving rubbish containers into storage or unpacking and installing a new piece of equipment. Others will be performing science experiments in the various laboratories on the space station.

12.30-1.00: Lunch

Crew members often eat lunch where they are working. Some astronauts exercise either side of lunch. All astronauts work out most days, once or twice a day, aiming for up to 2 ½ hours exercise.

16

Slowly Does It

Working in microgravity can be frustrating and takes patience. Tasks can take longer than on Earth because of having to plan every little detail, and having to secure every single object involved in the work otherwise it will float away. Astronauts use a lot of clips, sticky tape and Velcro fasteners to keep things in place.

1.00-6.00: More Work

The crew get back to working hard. Some might perform checks on key parts of the station, repair or replace air filters or fans, or prepare for a spacewalk or the arrival of another spacecraft the next day.

6.15-8.00: Early Evening

The crew contact mission control at the end of the working day to discuss how things went. Dinner is prepared and enjoyed usually by the crew all together. Dinner things must be cleaned away with the crew using a vacuum to hoover up any floating crumbs.

8.00-11.00: Pre-Sleep

Some of this time is spent planning for the next day's tasks, relaxing or watching a film. Astronauts may undergo some more medical checks. One or more astronauts may have private video conference calls with their family back on Earth or take part in question and answer sessions via live satellite links with schools and the public.

11.00: Bed Time

Phew! After all of that activity, it's no surprise that astronauts head to bed ready to sleep. Their bed, though, is a sleeping bag with holes for the arms to stick out, attached to the wall so that it won't float away.

Science In Space

The ISS is all about science. Thousands of science experiments have already been performed on the space station and many hundreds more are being devised and planned for future missions.

Experiment

Some of the science work performed on the ISS involves monitoring the Earth from space for signs of pollution and changing weather patterns. Other work uses scientific instruments mounted on the outside of the space station. Many experiments look at how living cells, whole plants and tiny living things, such as bacteria, develop in microgravity.

In The Glovebox

Many experiments are carried out on materials in space, learning how they burn or can be formed without the gravity found on Earth. Studying liquids and burning flames in space is difficult and dangerous unless the experiments take place inside a sealed mini lab called a glovebox.

Space Bots

Robot arms already perform many helpful tasks outside the ISS, but further robots are tested out for work inside the spacecraft including flying robot balls and a human-like robot called Robonaut 2. It has two arms and the ability to perform fiddly tasks like replacing parts.

You Are The Experiment

Whilst astronauts carry out many experiments, yet more experiments are carried out on them. ISS crews' eyes, brain, reactions and muscle strength are amongst many aspects of their health that are checked on and experimented with. Scientists are keen to learn more about how space affects people in advance of longer space missions possibly to the Moon or even Mars.

What's For Dinner?

Meals are planned long before astronauts head into space. Food is stored carefully to prevent it from going off. Some foods are canned and others are freeze-dried then sealed in pouches. Removing the water from food in this way not only helps preserve them, but also reduces their weight. Astronauts add water to these foods to re-hydrate them before eating.

Drink Up

Drinks are served in pouches too. These are equipped with built-in straws or nozzles so that the liquid cannot spill out. Some drinks are stored as powders in the pouch and have to have water added to them.

Dinner Table

A fold-out table provides a dining area for the crew but with them floating in space, there's no need for chairs. Metal trays are magnetised so that metal cutlery and scissors to open all the pouches stay in place. Food and drinks are strapped down using elastic cords or stuck to strips of Velcro.

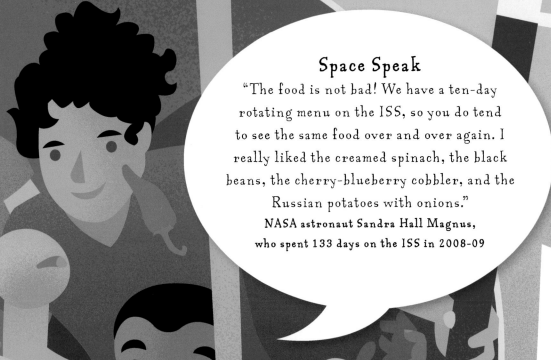

Space Speak

"The food is not bad! We have a ten-day rotating menu on the ISS, so you do tend to see the same food over and over again. I really liked the creamed spinach, the black beans, the cherry-blueberry cobbler, and the Russian potatoes with onions."
NASA astronaut Sandra Hall Magnus, who spent 133 days on the ISS in 2008-09

Taste In Space

Your sense of smell and taste isn't quite the same as on Earth. Some astronauts find chilli sauce or other spices help give their space meals more flavour. Salt and pepper are provided but in liquid bottles so that salt and pepper granules don't float away and clog up controls or instruments.

Fresh Food

Every few months, unmanned spacecraft bring new supplies including food to the space station. Astronauts get to enjoy some fresh foods like apples and vegetables. In 2015, the crew ate the first food grown entirely in space – red romaine lettuce – as part of a science experiment called Veg-01.

At Your Leisure

Astronauts work very hard so they enjoy the time they get to relax. They spend it in lots of different ways from enjoying the weightlessness in space to using an Internet link to stay in touch with family and friends.

Library Time

There's a small book and DVD library on the ISS. In addition, brand new films are sometimes uploaded to the astronauts' computers whilst live sports or entertainment events are sometimes shown on board, beamed by satellite to the space station.

A Little Bit Of Home

Every astronaut is allowed a small pack, weighing up to 1.5 kg, which they can fill with their own possessions for use on the ISS. Many take prized books or photos of their family. Japanese astronaut, Satoshi Furukawa, took a LEGO set and built a model of the ISS inside the space station in 2012.

Music Mission

In the third expedition to the ISS in 2001, commander Frank L. Culbertson decided he'd use his spare time to practise his trumpet playing! The trend for musical instruments has continued. NASA astronaut, Catherine 'Cady' Coleman brought not one but four flutes with her into space on her 2010-11 mission.

Chris Hadfield carried a three-quarter sized guitar to the ISS in 2012 and composed and recorded a whole album's worth of songs whilst in space. His version of David Bowie's 'Space Oddity' became an Internet hit, gaining 34 million views on YouTube. He also took part in a live concert across Canada with almost one million people singing along with him.

Space Speak

"Music makes it seem less like a spaceship, and more like a home." NASA astronaut Carl Walz, who played guitar on the ISS during his 196-day mission in 2003

Space Style, Space Hygiene

So how do you keep clean in space? There's no bath or shower on the space station, so astronauts clean themselves with towels made damp by adding a little soapy water from a sealed pouch or by using special wet wipes each sealed in foil and containing disinfectant.

Leave It In

Special rinseless shampoo is rubbed into hair and left there to keep it clean. This shampoo was originally developed for use by hospital patients on Earth who were unable to take a shower.

Brushing your teeth also presents problems as there's no sink to spit into. Instead astronauts must swallow their toothpaste or spit into a towel.

Packing For Space

There's no laundry or washing machine on the ISS so clothing is worn a lot longer than on Earth. T-shirts, shirts, trousers and shorts are changed on average once every 10 days. When clothing is finally deemed too dirty, it is usually treated as waste.

Space Speak

"If your body was clean, the clothes would stay clean for a long time because they basically float on you ... I wore the same pair of khaki shorts for three months and they were fine."

Nicole Stott, astronaut on two ISS missions, 20 and 21

Skip To The Loo

There are two loos on the International Space Station. Both use fan suction systems to create a flow of air that takes the place of gravity and sucks the waste away without flushing with water. Solid wastes is stored in sealed bags, which are compacted to take up less room. The bags are stored in aluminium containers before they are taken away from the ISS by an unmanned cargo spacecraft that will burn up in the Earth's atmosphere.

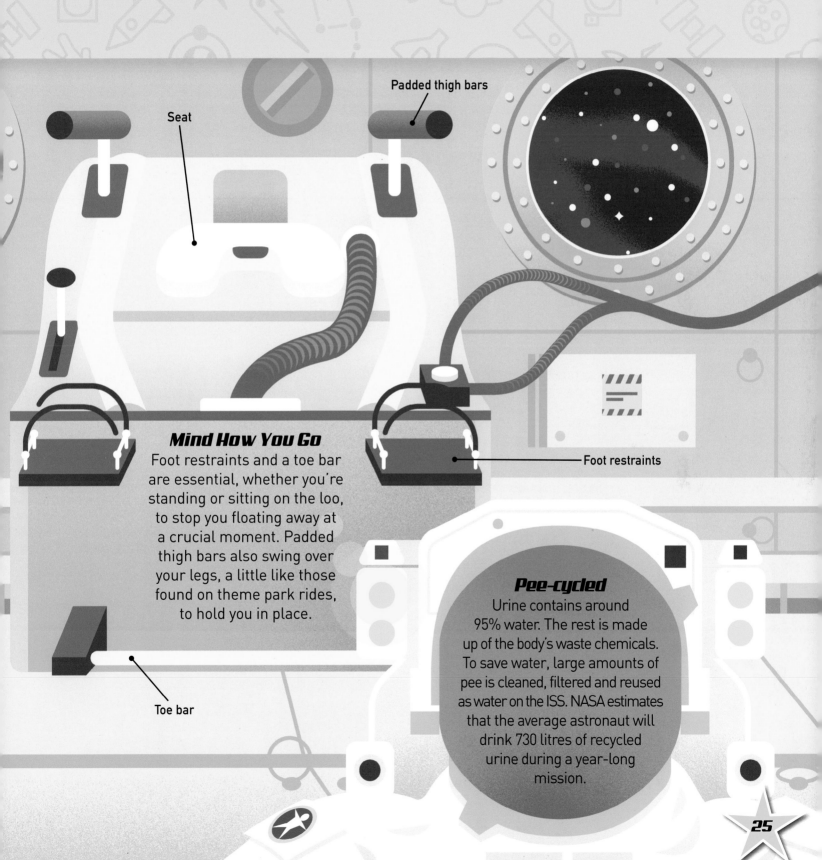

Seat

Padded thigh bars

Foot restraints

Mind How You Go

Foot restraints and a toe bar are essential, whether you're standing or sitting on the loo, to stop you floating away at a crucial moment. Padded thigh bars also swing over your legs, a little like those found on theme park rides, to hold you in place.

Toe bar

Pee-cycled

Urine contains around 95% water. The rest is made up of the body's waste chemicals. To save water, large amounts of pee is cleaned, filtered and reused as water on the ISS. NASA estimates that the average astronaut will drink 730 litres of recycled urine during a year-long mission.

Going Outside

Astronauts sometimes head outside the ISS to repair, or add new parts to, the space station. They may also go on a spacewalk to set up new science experiments or to check on existing ones. Scientists call these trips out extravehicular activities or EVAs for short.

Dangers of Space

Space is a hostile place for humans. There's no air to breathe and temperatures outside the space station can range from a terribly cold -129°C to a sizzling 121°C when in direct sunlight. In addition, there are harmful ultraviolet (UV) rays and tiny whizzing particles of rocks in space, which can rip through unprotected flesh.

Suiting Up

It takes over an hour to put on a spacesuit, which acts as a mini spacecraft, protecting the astronaut from the dangers of space. Plenty more time is spent performing dozens of safety checks before an astronaut can venture outside.

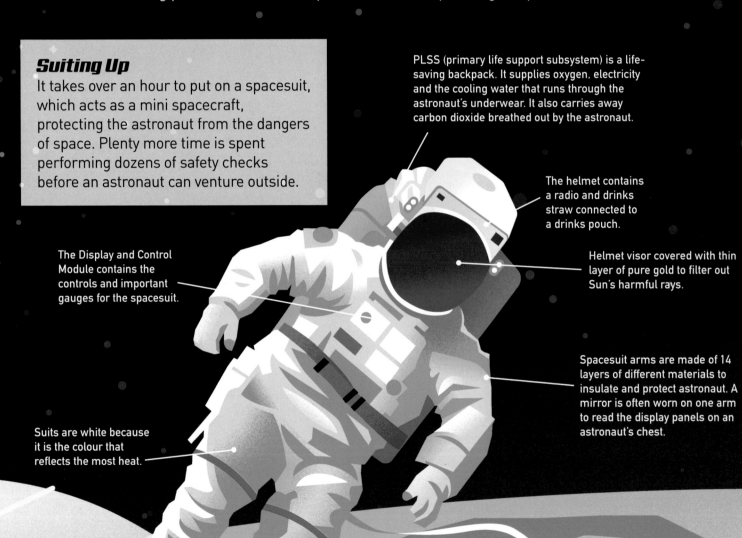

PLSS (primary life support subsystem) is a life-saving backpack. It supplies oxygen, electricity and the cooling water that runs through the astronaut's underwear. It also carries away carbon dioxide breathed out by the astronaut.

The helmet contains a radio and drinks straw connected to a drinks pouch.

Helmet visor covered with thin layer of pure gold to filter out Sun's harmful rays.

The Display and Control Module contains the controls and important gauges for the spacesuit.

Spacesuit arms are made of 14 layers of different materials to insulate and protect astronaut. A mirror is often worn on one arm to read the display panels on an astronaut's chest.

Suits are white because it is the colour that reflects the most heat.

In Your Pants

Underneath the outer suit, astronauts wear an electric harness, which carries all the wiring between the different parts of the suit and water-cooled underwear. Over 90 m of pipes weave through their underclothing. With spacewalks often lasting six hours or more and no loos in space, astronauts also wear a Maximum Absorption Garment – an adult nappy!

Getting Around

Once out in space, astronauts are tethered to the outside of the space station by cords so they cannot float away. Sometimes, the feet of their spacesuit are clamped to the end of the giant robotic arm which can move them more quickly.

Coming Home

All good things must come to an end, and eventually the day arrives when astronauts have to head back to Earth. Departing astronauts squash themselves into the tiny cramped Soyuz space capsule. After many safety checks have been made, the latches are released and their spacecraft starts to drift away from the space station. Once it is around 19–20 km from the ISS, the Soyuz fires its rocket engines for more than four minutes. This sends the spacecraft back towards Earth at 28,000 km/h.

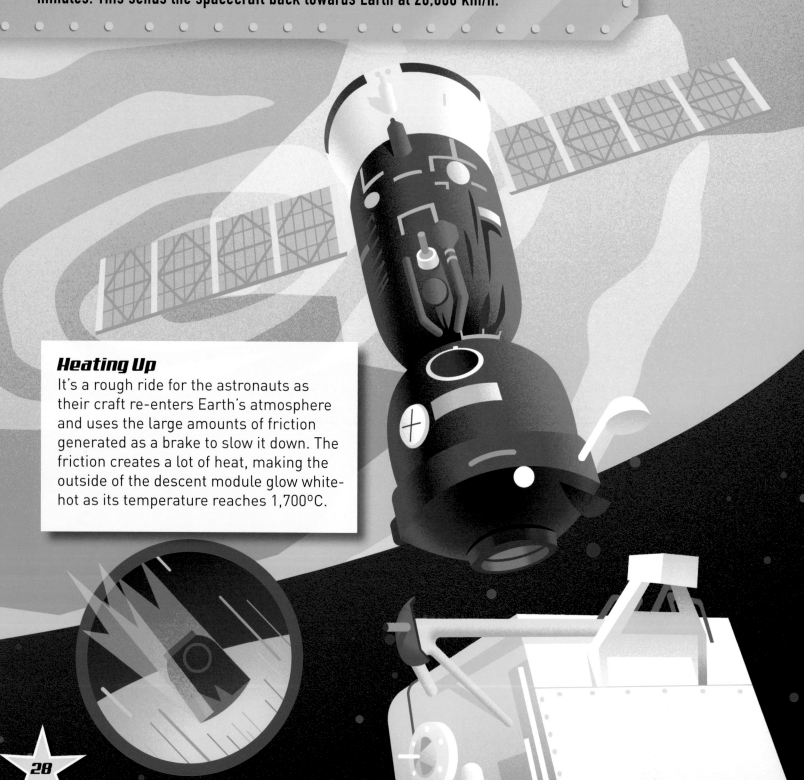

Heating Up

It's a rough ride for the astronauts as their craft re-enters Earth's atmosphere and uses the large amounts of friction generated as a brake to slow it down. The friction creates a lot of heat, making the outside of the descent module glow white-hot as its temperature reaches 1,700ºC.

Happy Landings

Once the craft has descended to 10,000 m above Earth, a small parachute is released pulling out a bigger chute and finally, a giant main chute. This slows the spacecraft down from 828 km/h down to 25 km/h. A second before impact on the ground, six small solid fuel rockets fire to cushion the landing.

Space Speak

"Right after I landed, I could feel the weight of my lips and tongue and I had to change how I was talking."

ISS astronaut, Chris Hadfield, days after he landed back on Earth

Back Down To Earth

Astronauts are looked after carefully in the weeks after their return from space. It takes them time to cope with Earth's gravity after a long period in the weightlessness of space. Some astronauts have to re-learn how to stand upright and walk in Earth's gravity and all train hard for months to regain their lost strength.

Glossary

Atmosphere The layer of gases surrounding the Earth.

Cell (living) The smallest unit of all living things, plants and animals.

CPR The abbreviation for pulmonary resuscitation: breathing air into the mouth of someone who is unconscious and pressing on their chest to keep them alive.

Docking port The part of a spacecraft where one spacecraft can join another in space.

Friction A force between two surfaces that are trying to move over each other, slowing them down.

Gravity The force that attracts objects towards each other. Earth's gravity keeps us on the ground.

Microgravity In space the force of gravity is so weak (microgravity) that weightlessness occurs.

Mission For spacecraft, the name for the special journey they make into space.

Mission control The place on Earth from which a space mission is controlled.

Module A detachable part of a spacecraft.

Orbit The path of a spacecraft, planet or star around another star or planet.

Radar A piece of equipment that uses radio waves to detect the presence of other objects.

Satellite link Transmitting messages via a communications satellite, an object that orbits Earth.

Space shuttle NASA's reusable spacecraft which were used to deliver crew, cargo and parts of the ISS in its early years.

Solar array A system of solar panels full of solar cells.

Solar cell A small device that converts the energy from the sun's rays into electrical energy.

Thermal radiator The part of the ISS that releases heat from the space station into space, to help regulate the temperature inside the station.

Time zones The Earth is divided into many equal parts, or time zones. In each area of land within a time zone, the time is the same.

Truss A long beam-like structure that joins something together.

Ultraviolet rays Invisible rays within sunlight that contain radiation, which can be harmful to humans.

Video conference call Using computers to provide a video link, callers can speak and see each other during a call.

Further Information

Websites

www.esa.int/esaKIDSen/index.html
Kids-themed webpages on the ISS produced by the European Space Agency.

www.nasa.gov/mission_pages/station/overview/index.html
NASA's reference guide to the ISS, packed with technical details, can be downloaded for free from this site.

http://iss.jaxa.jp/kids/en/life/
The Japanese Space Agency's website for kids is full of interesting facts about life in space, especially on the ISS.

https://spotthestation.nasa.gov/home.cfm
NASA's website allows you to find out when you can next see the space station in the night sky in your area of the world.

www.esa.int/Our_Activities/Human_Spaceflight/International_Space_Station/Highlights/
International_Space_Station_panoramic_tour
Get to see inside the ISS with these 360º panoramic photos of the different modules and labs, courtesy of the European Space Agency.

Books

Ground Control to Major Tim: The Space Adventures of Major Tim Peake by Clive Gifford (Wayland, 2017)

How To Design The World's Best Space Station by Paul Mason (Wayland, 2016)

Go Figure: A Maths Journey Through Space by Anne Rooney (Wayland, 2014)

The Story of Space Stations by Steve Parker (Franklin Watts, 2015)

Online videos to watch

http://www.space.com/18590-iss-tour-kitchen-bedrooms-the-latrine-video.html
Take four fascinating video tours of different parts of the space station with ISS astronaut, Sunita Williams, acting as your guide.

http://www.bbc.co.uk/newsround/36560356
Watch ISS astronaut, Tim Peake's top ten moments during his ISS mission.

https://www.nasa.gov/mission_pages/station/videos/index.html
A large collection of video clips to watch of ISS astronauts and the experiments they are working on.

Index